Comms For Preppers

M.Ray Davis

Published by Dragon Bear Publishing, LLC, 2025.

While every precaution has been taken in the preparation of this book, the publisher assumes no responsibility for errors or omissions, or for damages resulting from the use of the information contained herein.

COMMS FOR PREPPERS

First edition. September 8, 2025.

Copyright © 2025 M.Ray Davis.

ISBN: 979-8990677432

Written by M.Ray Davis.

Table of Contents

Introduction ... 1

Why Do Preppers Need Comms? ... 3

Comms 101 ... 5

Interlude ... 9

Citizens Band (CB) Radio ... 11

Family Radio Service (FRS) ... 13

Multi-Use Radio Service (MURS) ... 15

The Prepper Comms Plan ... 17

General Mobile Radio Service (GMRS) ... 23

Comms Initiative ... 27

Amateur Radio (Ham Radio) ... 31

The ARRL ... 41

The P.A.C.E. Plan ... 45

Wilderness Protocol: Is It Really Necessary? ... 49

#1000NewHams ... 55

Final Word ... 59

Appendix A ... 61

Appendix B ... 65

Appendix C ... 69

Appendix D 73

Appendix E 75

Appendix F 77

Final Thoughts 81

Acknowledgments

This book wouldn't exist without the prepping and ham radio communities—those who share their knowledge freely, test their gear in the field, and live the "be ready" mindset every day. To my fellow preppers and radio operators: thank you for keeping the spirit of resilience alive.

Introduction

I've been an amateur radio operator for about 13 years now, and one thing I've learned is this: when disaster strikes—before, during, or after—communication isn't just helpful, it's absolutely vital. You can stock up on food and water all day long, but if you can't reach out for help or coordinate with others, you're flying blind.

Now, amateur radio (a.k.a. "ham radio") takes some practice. You can't just pick up a transceiver and expect to know what you're doing—it's a skill, like driving stick shift or cooking a steak just right. And in the U.S., that skill comes with a license from the FCC. Why? Because if you mess around on the wrong frequencies without knowing what you're doing, you could end up with heavy fines, they might seize your equipment, and yes, in the worst cases—jail time. So, best to play by the rules. Wouldn't you agree?

That said, ham radio isn't the only game in town. There are other options like Citizens Band (CB), General Mobile Radio Service (GMRS), Family Radio Service (FRS), and Multi-Use Radio Service (MURS). The nice thing? FRS, CB, and MURS don't even require a license—you just buy the radio and go. GMRS and amateur radio do require a license, but honestly, it's worth looking into. Each of these services has a role to play, depending on your needs.

As for me—I've been a prepper since before the whole Y2K scare. Remember that? The world was supposed to grind to a halt at midnight when the computers forgot how to count past 1999. Didn't happen (thankfully), but the idea of being ready stuck with me. Back then, being a prepper meant people looked at you sideways—like you were paranoid or had a bunker full of beans and tinfoil hats. But really, we just wanted to be less dependent on "the system." If the power went out, we could keep living like normal. If we lost a job, the pantry was already stocked. It wasn't paranoia—it was insurance.

But as my prepping went on, I realized I had a gap: communication. Sure, I had food and gear, but how was I going to coordinate if things went bad? In 2007, the FCC dropped the old Morse code requirement for amateur radio licensing. Two years later, I took the plunge, got my license, and I haven't looked back since.

One last note before we dive deeper: I'm not here to push any politics. Doesn't matter which side of the aisle you're on—what matters is that you have the knowledge and tools to keep yourself and your people connected when it counts. That's all this is about. You and me, talking radios, so you can become a Radio Prepper too.

Why Do Preppers Need Comms?

When things go sideways, information becomes just as critical as food and water. In fact, for the prepping community, it's often the difference between staying ahead of the storm—or getting blindsided by it. We need to know what's happening not just in our own neighborhood, but in the next county over, the next state over, and yes—even across the globe. That flow of information in and out is what keeps us connected, coordinated, and calm.

That's where radios earn their keep. Having a $25 "Made in China" handheld stuffed in your bug-out bag is fine and all, but let me ask: do you actually know how to use it? Can you hit the national calling frequency on the 2-meter band without fumbling through the manual? How about the 70-centimeter band? Because when the lights are out and cell towers are silent, that little radio becomes your lifeline.

And it's not just about the hardware. It's about the plan. Do you and your group—whether you call them your MAG (Mutual Assistance Group), your team, or your tribe—actually have a communications plan? More importantly, have you practiced it? Writing down "We'll meet on Channel 3" is a start, but unless you run drills and make it second nature, chances are good that when the pressure's on, somebody's going to forget how to even turn the thing on.

That's why I'm writing this—to give you, my fellow preppers, the know-how to actually use your gear. To set up a comms plan that makes sense for your crew. And to do it all legally and effectively under U.S. rules, since that's the airspace I operate in. You'll see plenty of tips, frequencies, and practical advice ahead—but the real point is simple: communication is survival. Without it, you're just guessing in the dark.

Comms 101

Before you run out and buy a radio (or three), take a breath and ask yourself a few simple questions:

Who am I trying to contact?

How far away do I need to reach?

What communication tools do I/we already have?

What are the legal rules for the service I want to use (FRS, GMRS, CB, ham radio)?

And most importantly—what happens if I push that "push-to-talk" button without a license on a service that requires one?

The first question is usually straightforward. Odds are you're trying to reach family, friends, or your MAG (Mutual Assistance Group). That's the bread-and-butter stuff—"I'm okay," "We need help," or "Dinner's ready."

The second question—how far do I want to communicate—is where most preppers get tripped up. A lot of folks assume a Baofeng handheld is the holy grail of comms. Truth bomb: it's not. A dual-band handheld running on simplex (radio-to-radio, no repeaters) will usually get you about 3 miles with no obstructions. In the suburbs or the city? That range can drop in half thanks to all the buildings in the way.

Here's why: handhelds like these run FM, which is fantastic for clear audio and cutting through static—but walls, trees, and hills eat those signals alive.

Antennas Matter (A Lot)

That stubby little antenna your radio came with? Pure garbage. Do yourself a favor and upgrade to a decent aftermarket dual-band antenna (2-meter/70cm). Get the wrong one, and your radio's basically a paperweight.

For base stations, the higher the antenna, the better. Think above the treeline, not tucked in your basement window. Can't install an outdoor antenna? Get creative—a magnetic-mount antenna stuck on a pizza pan will do the trick in a pinch. (Yes, really—the pizza pan acts as a ground plane. No delivery required.)

Why People Love Baofengs

Baofengs are cheap, which makes them popular. But they're only really good for local, neighborhood-level communication—unless you can hit a repeater.

A repeater is basically a big electronic megaphone. It listens to your weak handheld signal, boosts it, and re-transmits it over a much larger area (20–30 miles or more). Programming them into your radio can be fiddly, but the free CHIRP software makes it way easier.

The catch? In an SHTF scenario, don't count on repeaters. They run on backup power, and when the fuel runs out, they go silent. So, your little handhelds are great for group comms, but they won't magically let you check in on Aunt Sue two states away.

That's where HF (High Frequency) radios come in. Unlike VHF/UHF handhelds, HF doesn't rely on repeaters. With the right antenna and a little skill, you can talk regionally—or even worldwide. (We'll dig into that more later.)

Don't Forget What You Already Have

Right now, you probably already own a comms device: your cell phone. In a disaster, though, cell service is shaky at best. Towers get overloaded fast. If you can't make a call, try a text—it uses less bandwidth and has a better chance of sneaking through. (Think of it as the Morse Code of the cell phone world: simple, narrow, effective.)

Power and Safety

The reach of your signal isn't just about the radio—it's about power and antenna. More watts = more punch. Most handhelds run 5–8 watts, which is plenty for local use. If you want more power, step up to a mobile radio (25–65 watts) and run it as a base station.

And don't stress too much about RF (radio frequency) exposure. According to the American Cancer Society, RF energy is non-ionizing—it doesn't knock electrons out of atoms. In crazy high amounts, it can cause burns, but your average prepper radio isn't cooking anyone.

Antenna Options in Plain English

Handheld "rubber duck" antennas: work, but barely. Upgrade ASAP.

Mag-mount antennas: perfect for slapping on a car roof—or a pizza pan in your apartment.

Base station antennas: taller is better. Copper J-Poles are a favorite—buy one or build your own.

Distance Reality Check

Baofeng UV-5R simplex: 1–5 miles (best case).

With a repeater: 30–50 miles, sometimes more if they're linked.

HF radio: regional to worldwide, depending on the band and conditions.

HF: Playing the Long Game

HF radios don't need repeaters. Just you, your radio, your antenna, and power. Some bands are better in the daytime (like 20 meters), others work best at night (like 40 meters). Learn when to use what, and you can reach out far beyond your county lines.

Bottom line: Radios are like any other prep—you don't just buy them and toss them on a shelf. You practice. You learn. You make mistakes now, so you don't make them when it really matters. Because in comms, just like with food, water, and shelter, knowing how to use your gear is half the battle.

Interlude

I'm an Amateur Radio Operator, and I'll be honest—I love the challenge of working QRP (that's low power, for the uninitiated). There's something exhilarating about bouncing a tiny 5-watt signal off the atmosphere and making contact hundreds—or even thousands—of miles away. I've also poured a lot of energy into encouraging folks to join the ham ranks, even pushing things like the #1000NewHams initiative.

But here's the thing: along the way, I realized I may have left some people behind. Not everyone wants—or is able—to get their Amateur Radio license. Maybe they don't have the time to study. Maybe test anxiety is real. Maybe they just don't see themselves diving into the technical side of ham radio. And you know what? That's okay. Communication shouldn't be an exclusive club.

If you're one of those folks, I don't want you walking away thinking, "Well, I guess radios aren't for me." Because they are. There are license-free options that still give you valuable comms capability. Will they make you the next worldwide DX champion? Nope. But for local, practical prepping purposes—they're absolutely viable.

Citizens Band (CB) Radio

If radios were high school, CB would be the kid who peaked in the 1970s but still has some tricks up his sleeve. Born in the post-war boom and immortalized in trucker movies and country songs, CB has been around long enough that almost everyone has heard of it—even if they've never used it.

How it works:

CB operates on 40 channels in the 27 MHz band.

Maximum legal power: 4 watts AM or 12 watts SSB (sideband).

No license required in the U.S.—just buy it, plug it in, and talk.

Pros for preppers:

Dirt cheap and widely available—pawn shops, truck stops, Amazon, you name it.

Excellent for vehicle-to-vehicle comms in a convoy.

Sideband (SSB) gives you more range than standard AM.

Big enough user base that you can always find chatter, especially from truckers.

Cons:

Crowded and noisy, with plenty of "colorful characters" hogging the channels.

Limited range—usually a few miles unless conditions are just right.

Big antennas required for best performance (you're not hiding one of these in your back pocket).

Prepper angle:

CB is perfect for local intel gathering. Truckers are mobile scouts—they'll tell you about accidents, fuel availability, and roadblocks long before your news app updates (assuming your news app even works). It's also handy for rural groups spread across a few miles of farmland. Don't expect crystal-clear tactical whispering, but do expect solid situational awareness.

Family Radio Service (FRS)

FRS is the "grab-and-go" of radios. These are the little blister-pack walkie-talkies you see at Walmart or sporting goods stores with marketing promises like "Up to 30 miles of range!" Spoiler: unless you're standing on two mountaintops with perfect line-of-sight, you're not getting anywhere close to that.

How it works:

22 channels in the UHF band.

Power capped at 2 watts.

Fixed antennas—no swapping in upgrades.

100% license-free.

Pros for preppers:

Cheap, simple, and readily available.

No license hoops to jump through—just hand them out and start talking.

Perfect for short-range, close-in coordination.

Small and lightweight, easy to stash in a go-bag.

Cons:

Range is limited—realistically 1–2 miles in suburban terrain.

No external antennas means no way to boost performance.

Everyone and their cousin owns a set, which means channels can get crowded.

Prepper angle:

Think of FRS as your intra-squad comms. Perfect for keeping family members in touch during a camping trip, neighborhood patrol, or even just running around a property. They shine in short-range coordination: "Meet back at the rally point," "Dinner's ready," or "Stay put until I get there." They won't bring you breaking news from across the county, but they will keep your immediate people connected.

Multi-Use Radio Service (MURS)

MURS is the best-kept secret in license-free comms. It doesn't get the spotlight like CB or the shelf space like FRS, but for preppers, it's a hidden gem.

How it works:

5 channels in the VHF band (around 151–154 MHz).

Power capped at 2 watts.

Unlike FRS, you can use external antennas (this is the game-changer).

License-free in the U.S.

Pros for preppers:

Less crowded than FRS or CB—fewer people even know it exists.

External antennas = better range and performance.

Great balance between portability and utility.

Solid for property security or neighborhood-level comms.

Cons:

Limited selection of radios—fewer options on the market.

Lower power than CB, so range is still modest.

Because it's lesser-known, finding compatible gear for everyone in your group can be a hurdle.

Prepper angle:

MURS is excellent for property defense and short-range ops. Think farm-to-farm, ranch security, or neighborhood patrol. Because it flies under the radar (pun intended), you're less likely to get background noise from non-preppers. And with the ability to add an external antenna, you can squeeze a lot more performance out of MURS than you ever could from an FRS handheld.

The Bottom Line on License-Free Radios

If you don't want to (or can't) get an amateur license, you're not out of the game. CB, FRS, and MURS are all plug-and-play solutions that can give you real, tangible comms capability when the lights go out.

CB: Great for road intel and local town-to-town comms.

FRS: Perfect for family and close-range team use.

MURS: The stealthy workhorse for property-level security.

None of them are perfect. All of them are limited. But in a grid-down situation, some communication beats no communication every single time.

The Prepper Comms Plan

If you take only one thing away from this chapter, let it be this: you need a comms plan before disaster strikes. Gear is good, but without a plan, your radios are just expensive paperweights.

A comms plan is nothing more than a pre-arranged agreement between you and your group (your MAG, family, or tribe) about how you'll communicate when the grid is down. It doesn't have to be complicated, but it does have to be written, shared, and practiced.

Here's what every comms plan should cover:

Who are we talking to?

Family? Neighbors? MAG members? Wider community? Know your core circle first.

How are we talking?

Decide what tools you'll use: ham radio, GMRS, FRS, CB, MURS, or even cell phones (if they're still working).

When are we talking?

Set specific times to monitor or transmit—like 10 minutes at the top of every hour. That way, even if you can't be glued to your radio 24/7, you're still in sync.

Does everyone know the plan?

Every member of your group should have the same info: programmed frequencies, call signs (if licensed), codes or brevity signals, and scheduled check-in times.

My Personal Example

For me, amateur (ham) radio is my backbone. I monitor the national 2-meter calling frequency, 146.520 MHz, for a few minutes every hour. At the same time, I keep an ear on Channel 3 across CB, FRS, and MURS—because you never know what might come through.

I also check into local 2-meter nets through repeaters as part of my county's ARES (Amateur Radio Emergency Service). This keeps me sharp, teaches me net etiquette, and makes sure my gear is always dialed in.

Oh, and yes—I keep my cell phone handy too. Even in a storm or power outage, sometimes text messages still sneak through when calls won't. A quick "We're safe" text to out-of-town family can save a lot of worry.

Building Your Plan

If your group decides to use ham radio, you'll need to pick a simplex frequency (radio-to-radio) for local comms, and also know what repeaters are available in your area. Every repeater has:

A receive frequency (what you listen on).

A transmit frequency (what you send on).

A shift (+ or –).

And possibly a tone (called a CTCSS tone).

All of that info is easy to find on RadioReference.com or RepeaterBook.com. Write it down. Program it in. Practice with it.

If you're sticking to license-free options like CB, FRS, or MURS, the same logic applies—pick your channels, set your schedules, and get everyone on the same page.

Nets: Practice in Action

In the ham world, we have something called nets—on-air gatherings that meet at a set time, on a set frequency, for a specific purpose. There are two flavors:

Formal (directed) nets: run by a Net Control Station (NCS). They follow a structured order: emergency traffic first, then check-ins, then a discussion topic. These are fantastic for emergency coordination.

Informal (round-robin) nets: more relaxed. Everyone gets a turn to share. Great for practice, camaraderie, or just checking your gear.

And here's the cool part: nets aren't just for ham radio. You can set up your own local net on FRS, CB, or MURS. Even a casual weekly check-in gives your group practice and builds confidence. Plus, it's fun—it turns "practice" into a social thing.

The Why Behind It All

When disaster hits, you must know what's happening around you. Is the storm shifting direction? Did the bridge wash out? Are emergency services setting up shelters? A simple battery-powered radio can bring in local news. But to push information out—to let your group know you're safe, to call for help, to share intel—you need a comms plan and the discipline to use it.

Part 2: GMRS and Amateur Radio

Introduction

I've been an Amateur Radio operator for over 13 years now, and one thing has become crystal clear: communication before, during, and after a disaster isn't optional—it's survival. Food, water, and shelter will always be top priorities, but without the ability to pass and receive information, you're essentially operating blind.

Here's the catch: you can't just pick up an Amateur Radio transceiver and start broadcasting. In the United States, the Federal Communications Commission (FCC) regulates the airwaves. To practice legally, you need a license grant. Operating without one isn't just bad form—it's expensive. We're talking hefty fines, confiscated gear, and yes, even potential jail time. Staying on the right side of the law isn't just about avoiding penalties; it's about keeping the radio spectrum useful and reliable for everyone.

That said, Amateur Radio isn't the only option out there. The airwaves are divided into different services—CB, GMRS, FRS, MURS, and Amateur (ham) Radio. Some, like CB, FRS, and MURS, are "license by rule." That means you can buy a radio, turn it on, and start using it without any paperwork. Others, like GMRS and Amateur Radio, require an actual FCC license. Each service has its place, and for preppers, the more you understand your options, the more resilient your comms plan becomes.

My Prepping Journey into Radio

I've been in the prepping world since before Y2K—remember that one? We were told every computer system on Earth was going to roll over to

"00" and collapse civilization. The apocalypse didn't show up that night, but prepping stuck with me.

Back then, being a prepper was seen as fringe, maybe even paranoid. People looked at us like we were weirdos with basements full of beans. Today, preparedness is far more mainstream, and the logic is simple: we choose to take responsibility for our own well-being. We stock food, water, and medical supplies. We invest in alternative power. We think ahead about job loss, weather events, or grid outages—not out of fear, but because it just makes sense.

But even with my shelves stocked and my gear squared away, I knew there was a gap in my plan: communication. What good is all the rice and first aid kits if you don't know what's happening outside your front door?

That's when Amateur Radio caught my attention. In 2007, the FCC dropped the Morse Code (CW) requirement, which had scared off a lot of potential hams. Two years later, I finally went for it, got licensed, and it completely changed my prepping game. Having that little card in my wallet meant I could learn, practice, and legally operate equipment that opened doors far beyond CB or FRS. It turned my radio from a toy into a tool.

A Note on Politics

Before we dive deeper, let me be clear: I'm keeping this book as politically neutral as possible. Prepping—and especially radio prepping—isn't about red, blue, or anything in between. It's about making sure your family, your group, and your community can stay connected when the grid goes down. That's it. No matter where you fall on the spectrum, the airwaves are for everyone.

My only goal here is simple: to arm you, Dear Reader, with the knowledge and confidence to step into the role of a Radio Prepper. Whether you choose to explore GMRS, commit to getting your ham license, or build a layered approach that uses multiple services, this is about empowerment.

General Mobile Radio Service (GMRS)

If FRS is the "starter walkie-talkie" for families, then GMRS is its older, tougher sibling. It looks similar on the surface, but under the hood, it's a completely different animal—and for preppers, that makes all the difference.

A Little Background

GMRS has been around since the 1980s, designed originally for families and small groups to have a more capable option than FRS or CB. Unlike Amateur Radio, GMRS doesn't require a test—but it does require a license. The license is granted by the FCC, costs about $35 (as of this writing), and lasts for ten years. Here's the kicker: one license covers your entire family. That means you, your spouse, and your kids can all legally use GMRS under one umbrella.

How GMRS Works

Frequencies: 22 channels in the UHF band (462–467 MHz). These overlap with FRS, but GMRS radios are allowed higher power and external antennas.

Power: Up to 50 watts on certain channels. (Compare that to FRS's measly 2-watt cap.)

Antennas: Unlike FRS, GMRS radios can use detachable or external antennas, which is a massive upgrade in performance.

Repeaters: Here's the game-changer. GMRS allows the use of repeaters—just like Amateur Radio. A well-placed GMRS repeater can expand your range from a few miles to dozens.

Pros for Preppers

Simple to get started. No test, just a fee and a license.

Family coverage. One license = everyone in your household is legal.

More power. 5 watts handheld? Sure. 25–50 watts mobile or base? Absolutely. That extra punch matters when you need to push a signal through trees, hills, or buildings.

Repeaters. Access to GMRS repeaters (and yes, they're popping up more and more across the U.S.) can turn a local radio into a regional tool.

Better gear. Because GMRS is growing in popularity, companies like Midland, Wouxun, and others are making solid, affordable radios specifically for this service.

Cons

License required. It's not free like CB, FRS, or MURS—you've got to register with the FCC.

Limited compared to ham. While GMRS is powerful, it doesn't give you the worldwide reach or flexibility of Amateur Radio.

Dependent on repeaters. Just like ham, if you're relying on repeaters, keep in mind they may not survive grid-down scenarios.

The Prepper Angle

GMRS shines as a family and group communication tool. Think convoy coordination, property security, or neighborhood patrols. The ability to push 25–50 watts through a mobile radio with an external antenna means you can realistically cover your town or rural county. Add a repeater, and your reach might extend to the next county over.

For groups who aren't ready to dive into Amateur Radio, GMRS is a fantastic stepping stone. It gives you legal power, repeater access, and enough capability to build a solid communications layer without requiring anyone to sit for a test.

Comms Initiative

This comes from AmRRON, the American Redoubt Radio Operators Network. This could be a great tool in our Prepping tool kit as it combines both the non-licensed communication options as well as the licensed.

A 'Net' is a regularly-scheduled communications network and is a great way to practice, become proficient with your equipment, and make contact with like-minded preppers and patriots in your area.

The reasoning behind this initiative is to get more people using the communications equipment that they have on hand. Communication during an SHTF Situation is paramount for survival, you will need to know what is happening, where it's happening, and who is involved.

If you only have a CB Radio, then you can use it to gather and disseminate information. True it is s short-distance communication device, but with the proper antenna height, you can receive and transmit signals just fine.

Ham Radio can be used for a longer range, and the Ham Ops should have a CB unit to monitor ch 3 in order to relay any information that comes via that avenue on the Ham Radio Frequencies locally, regionally, and worldwide.

Ham Radio is licensed by the FCC here in the US, and before you say "But in an SHTF Situation licenses will not matter" think about this, would you pick up a firearm that you've never practiced with and be able to be an expert in its use? No. Ham Radio is no different, these are not "Plug and Play" items, and you have to actually know how to use them. That comes with practice and getting a license is the best way to start. I am well aware that some in the preparedness community do not want to "Be on a government list", here's a knowledge bomb for you, you are on a government list if 1) you own your own home, 2) Own a vehicle of any kind 3) have a Social Security Card. You were placed on a government list on the day you came into this world kicking and screaming.

Getting a license means that you have taken the time to learn how to use this type of equipment, just like driving a car.

Ham Radio is a way to "stay in the loop" before, during, and after an SHTF situation. When Mother Nature sends storms to your area Cell towers may very well be down, and power could be out but Ham radio can be operated via a battery, so it is a great option to have available to you.

Ham Radio does have some requirements, like needing a license to operate the radio. To get a license requires taking a 35 question, multiple-choice test. You can take a practice test at the ARRL website, or QRZ.Com for free until you consistently score at least 80%. The next step would be to find a local Ham Radio Club, as they generally offer the Technician (the first license for Amateur Radio in the USA) test at least monthly.

The License is good for 10 years, so even at $35, it breaks down to like $3.50 a day. As far as a "hobby" goes, that ain't too bad.

You may have noticed that I put the word hobby in quotes because some people only see what we do as a hobby. In fact, we are more than that, we provide service in times of crisis. During Hurricane Katrina, there were Ham Radio ops in and around New Orleans relaying health and welfare information out of the affected area. When Hurricanes Irma and Maria blasted Puerto Rico, Ham Radio Ops went down to establish communications for the whole Island. And this was done all without getting paid one red cent.

Now think about it this way, should an SHTF situation come, would you want to be able to communicate with other, like-minded people in order to coordinate relief efforts for yourself/MAG/TRIBE or Group? I know I would want a way to at least gather information, not necessarily INTEL, but information as to what is going on.

To start with, you do not need to get the biggest and best gear, quite frankly because it costs way too much.

Most preppers I know have started out with a Baofeng UV-5R This is a dual-band Amateur radio, meaning that it is capable of transmitting and receiving on the 2 Meter band (144.0 to 146.8 MHz) and on the 70 Centimeter Band (420.0 to 450.0 MHz) these are the Technician level authorized bands. For $25.00 on Amazon, you cannot go wrong.

Included in the Tech License are limited privileges on the 10 Meter band (28.300 to 28.500) for Single Side Band phone (voice) operation Anytone makes a good 10-meter radio for a decent price, see below for pic:

The 10 Meter band voice privileges open up continental, if not world wide communication for Preppers who have decided to become amateur radio operators..

That is the whole gist of this initiative, to get both forms of comms, Amateur, and CB to work together. By monitoring a certain channel on CB and on the Ham Bands an Operator can pass along health and wellness information, as well as information on conditions in your area to those outside your area during an SHTF situation.

Amateur Radio (Ham Radio)

What It Is and Why It Matters

Amateur Radio—better known as "Ham Radio"—isn't just a hobby, it's a global community. The amateur and amateur-satellite services exist for people who want to explore radio technique for personal growth, self-training, and public service—not for profit. It's about learning, experimenting, and connecting.

Worldwide, millions of hams communicate across 29 internationally recognized bands using every mode you can think of—voice, digital, Morse code, TV, even satellites. Unlike commercial radio, no amateur frequency is reserved for one person's exclusive use. We all share, cooperate, and coordinate to keep the airwaves useful.

For preppers, Ham Radio is the crown jewel of communications. It gives you tools to reach local, regional, and even worldwide contacts—something CB, FRS, GMRS, and MURS simply can't match. But there's a catch: to transmit legally on amateur frequencies, you need an FCC license. No license, no transmitting—period. Try it anyway and you risk big fines, losing your gear, and possibly jail time.

So yes, it's worth doing it the right way.

The Phonetic Alphabet

One of the first skills every ham learns is the phonetic alphabet. Why? Because in noisy conditions, some letters sound alike. Try telling the difference between "B" and "D" over static, and you'll see why we spell things out with words.

Here's the ITU Phonetic Alphabet (used worldwide):

A – Alpha
B – Bravo
C – Charlie
D – Delta
E – Echo
F – Foxtrot
G – Golf
H – Hotel
I – India
J – Juliet
K – Kilo
L – Lima
M – Mike

N – November
O – Oscar
P – Papa
Q – Quebec
R – Romeo
S – Sierra
T – Tango
U – Uniform
V – Victor
W – Whiskey
X – X-Ray
Y – Yankee
Z – Zulu

Pro tip: practice spelling your name, your call sign, and your group's name phonetically until it becomes second nature.

License Levels in the U.S.

There are three license classes here in the United States. Each step up gives you more privileges and more frequencies to work with:

Technician Class

Entry-level license.

35-question test on rules, safety, and basic radio theory.

Grants access to all amateur frequencies above 30 MHz (like 2 meters and 70 centimeters). Great for local VHF/UHF comms.

General Class

Mid-level license.

Another 35-question test. Must already hold a Technician license.

Grants operating privileges on most HF bands, which means you can finally reach nationwide and worldwide contacts.

Amateur Extra Class

The top tier.

50-question test covering advanced theory, regulations, and operating practices.

Full access to all amateur bands and modes. Nothing is off-limits.

Exams are given by Volunteer Examiners (VEs) in your community, organized by Volunteer Examiner Coordinators (VECs). The test questions come from published pools—so you can literally study the exact questions you'll see on the exam.

How to Study and Take the Exam

Back in the day, you had to hunt down a local ham club to test. Now you can do it face-to-face or online. Study guides, apps, and practice exams are everywhere:

HamStudy.org – Free practice tests and flashcards

HamRadio Prep – Courses and free quizzes (https://hamradioprep.com/free-ham-radio-practice-tests/)

ARRL Exam Review – Official test prep (http://arrlexamreview.appspot.com/)

QRZ.com – Practice tests and study forums

And if you want to register for an exam (online or in person), head over to the ARRL website:

http://www.arrl.org/online-exam-session

Once you pass, the VE team submits your info to the FCC through the Universal Licensing System (ULS). Your license usually shows up in the FCC database within a week, and from that moment on—you're legal.

Frequencies and Capabilities

Depending on your license class, amateur frequencies stretch from 160 meters (1.8 MHz) all the way up to 23 centimeters (1.2 GHz). That covers everything from local VHF/UHF (think line-of-sight, neighborhood-to-town) to worldwide HF (shortwave-style long-haul).

VHF/UHF (2m & 70cm): Local and regional, great for repeaters.

HF (40m, 20m, etc.): Nationwide and worldwide, depending on time of day and solar conditions.

Satellites: Yes, hams bounce signals off satellites. It's a thing.

Gear Costs Today

When I started, radios were expensive. A basic 2-meter handheld (HT) would run $200+. These days, thanks to brands like Baofeng, you can grab a dual-band handheld on Amazon for about $25. They're not perfect, but they're rugged, clear, and they work. With one of those, I can hit a local repeater from inside a brick building without a hitch.

Handhelds usually put out 5–8 watts, good for about 1–5 miles simplex (radio-to-radio). Add a repeater into the mix, and suddenly you're covering 20–50 miles or more. Step up to a mobile or base station radio (25–65 watts), and your reach grows even more—especially if you invest in a good antenna.

Remember: with VHF and UHF, line of sight is king. If your antenna can "see" the other guy's antenna, you're golden. Trees, hills, and buildings? Those eat signals fast. That's why antenna height often matters more than raw power.

Why It Matters for Preppers

In a crisis, knowing what's happening outside your immediate area is priceless. Weather alerts, road closures, shelter locations, and health/welfare messages all flow across amateur bands when the grid is down. And unlike CB, FRS, or even GMRS, Ham Radio gives you access to worldwide communication.

More importantly, ham operators train constantly. Weekly nets, emergency drills, and disaster response groups like ARES (Amateur Radio Emergency Service) keep skills sharp. As a prepper, tying into that network means you're not alone—you're part of a global web of radio operators ready to share information when the lights go out.

Getting Started in Ham Radio: A Step-by-Step Guide

So you've decided Amateur Radio might be worth your time. Good call. Now let's break down the process into bite-sized steps so it doesn't feel overwhelming.

Step 1: Learn the Basics

Don't panic—you don't need to be an engineer. The entry-level Technician license test is 35 multiple-choice questions. Topics include:

Basic radio theory (how signals travel)

Safety (like not zapping yourself with RF)

FCC rules and operating practices

Most folks can study for it in a few weeks of casual effort.

Step 2: Pick a Study Resource

You've got options:

HamStudy.org (free, flashcards + practice tests)

Ham Radio Prep (structured courses, good for beginners)

ARRL Exam Review (official study site)

YouTube – search "Technician ham radio crash course" for video lessons

Books – The ARRL Ham Radio License Manual is the gold standard

Tip: Take practice exams until you're consistently scoring above 80%. At that point, you're ready.

Step 3: Find an Exam Session

Exams are given by teams of Volunteer Examiners (VEs). You can take the test two ways:

In person: Usually hosted by ham radio clubs, often once a month.

Online: Thanks to Zoom and webcams, you can now test from home.

Find sessions at:

http://www.arrl.org/find-an-amateur-radio-license-exam-session

http://www.arrl.org/online-exam-session

Cost: usually $15 or less.

Step 4: Pass the Test

When you pass, the VE team sends your info to the FCC. Within about a week, your callsign will show up in the FCC database. That's your official green light to transmit. Print it, screenshot it, tattoo it (okay, maybe not that last one).

Congratulations—you're now a ham.

Step 5: Get Your First Radio

Start small, start simple:

Handheld (HT): A Baofeng UV-5R (about $25) is cheap and works. Not the Cadillac of radios, but great for learning.

Upgrade Antenna: Toss the stock "rubber duck" antenna and get an aftermarket dual-band whip. Huge difference in performance.

Accessories: Extra batteries, a simple programming cable, and CHIRP software (free) will make your life easier.

If you've got the budget, consider a mobile radio (25–50 watts) with a mag-mount antenna for your vehicle or even as a small base station at home.

Step 6: Program and Practice

Learn how to program your local repeaters into your radio (RepeaterBook.com is your friend).

Get on the air—listen first, then key up when you're confident.

Join a local net to practice. Nets are like on-air meetings; they're structured, friendly, and designed for exactly this purpose.

Step 7: Level Up

Once you're comfortable as a Technician:

Study for your General license (another 35 questions). This opens up HF bands and worldwide communication.

Eventually, shoot for Extra if you want full privileges.

Step 8: Connect with the Community

Ham radio is best learned with others. Join a local club, attend field days, and meet people who can help you grow. Most hams love teaching newcomers—it's part of the culture.

Step 9: Integrate Into Your Preps

Finally, make Amateur Radio part of your overall plan:

Write your frequencies into your Comms Plan.

Train your family or MAG on how to use the radios.

Run regular drills.

Keep backup power (extra batteries, solar chargers) handy.

Radios aren't "buy once and forget" gear. They're tools you sharpen through use.

Bottom Line

Getting into ham radio isn't as hard—or as expensive—as it used to be. For the cost of a tank of gas and a couple evenings of study, you can earn a license, buy a radio, and plug into a worldwide network of operators.

More importantly, you'll give yourself and your group a reliable way to communicate when everything else goes dark.

The ARRL

If you've spent more than five minutes in the ham world, you've probably heard of the ARRL—the American Radio Relay League. Founded on April 6, 1914, it's the largest and oldest amateur radio association in the United States. With over 161,000 members, the ARRL is basically the beating heart of the U.S. amateur radio community.

What the ARRL Does

The League wears a lot of hats. At its core, it exists to represent the interests of amateur operators before Congress and the FCC. When new laws or regulations are proposed that affect ham radio, the ARRL is the one standing up to say, "Hold on a second—here's what this means for the people actually using the airwaves." In other words, they're our lobbyists, our advocates, and our line of defense.

But it doesn't stop there. The ARRL also:

Provides technical assistance to members. Got a question about antennas, digital modes, or why your signal sounds like a dying robot? They've got experts for that.

Runs educational programs to introduce newcomers to the hobby and help veterans sharpen their skills.

Supports emergency communications nationwide through programs like ARES (Amateur Radio Emergency Service). This is a big deal for preppers, because it means the ARRL isn't just about fun—it's about real-world readiness when the grid goes dark.

Personally, that's one of the reasons I joined. I wanted to learn how to handle health and welfare traffic—the kind of messages people desperately need to get through when normal lines of communication fail.

Member Perks

Membership in the ARRL comes with a laundry list of benefits. A few highlights:

QSL Bureaus – If you're into DXing (long-distance contacts), you'll love this. The ARRL runs incoming and outgoing QSL bureaus so you can swap those classic contact cards with operators all over the world. There's something special about holding a postcard from halfway across the globe that says, "Yep, we talked."

Volunteer Examiner Coordinator (VEC) – The ARRL sponsors the exam system for all three U.S. license classes. I eventually became a Volunteer Examiner (VE) myself so I could help usher in new hams. It's a way of giving back to the community.

Contests – If competition is your thing, the ARRL sponsors some of the biggest amateur radio contests in the world, including the November Sweepstakes and the International DX Contest. These aren't just bragging-rights events—they push your skills, test your gear, and connect you with operators you'd never meet otherwise.

Publications & Resources – Membership includes subscriptions to QST magazine (packed with projects, tips, and news), plus access to tons of online resources.

Why Membership Matters

Look, you don't have to be an ARRL member to be a ham. But here's the way I see it:

They educate.

They connect.

They protect.

For the cost of a yearly membership, you're not just buying magazines and QSL services—you're helping fund the organization that fights to keep amateur radio alive and relevant in a world that's obsessed with Wi-Fi and smartphones.

As preppers, we understand the importance of having a voice—literally and figuratively. By joining the ARRL, you're making sure that voice continues to be heard in Washington and beyond. In my humble opinion, that alone makes membership worth it.

The P.A.C.E. Plan

When it comes to communications, one method is never enough. Batteries die, networks fail, gear breaks. That's why the military developed the P.A.C.E. Plan—a simple way to build redundancy into your communications (and honestly, it works just as well for any other critical system).

P.A.C.E. stands for:

Primary

Alternate

Contingency

Emergency

Think of it as your layered safety net. If one option fails, you immediately fall back to the next. Let's break it down.

Primary – Everyday Use

For most of us, the smartphone is the primary comms tool. Voice calls, text messages, apps—it's quick, familiar, and effective. The problem? Cell networks are notoriously fragile. In disasters, towers overload or go offline completely.

That doesn't mean you ditch your phone. It just means you prep for its limitations. Keep it charged with:

A solar charger (great for off-grid or outdoor use).

A power bank—the kind that can recharge your phone multiple times.

A car adapter if your vehicle still runs.

Personally, I keep a solar power pack I snagged off Amazon for about $40. It's paid for itself during power outages more than once.

Alternate – Radio, Radio, Radio

When the cell network goes dark, it's time to lean on radios. My go-to is a handheld dual-band amateur transceiver (2m/70cm). With a Technician license, I can legally operate and reach out on frequencies like 146.520 MHz (the national 2-meter calling frequency).

Now, let's keep it real: handhelds aren't miracle machines. On flat, unobstructed terrain you might get ~3 miles simplex (radio-to-radio). Up on a ridge or mountain? You might do much better. Add a repeater into the mix, and suddenly you're covering counties instead of blocks.

The key is: radios don't depend on infrastructure in the same way cell towers do. If you've got a working radio and a charged battery, you've still got a voice.

Contingency – Advanced Tools

This is where you bring in your specialized systems. For me, that's APRS (Automatic Packet Reporting System).

APRS lets hams send digital data—like position, altitude, movement speed, or short messages—over radio. With something as simple as a Baofeng UV-5R, an audio interface cable, and an old smartphone, you've got yourself a portable APRS station. Suddenly, you're not just talking—you're sending tactical info your group can actually use.

Don't have APRS? No problem. GMRS also fits beautifully here if your family is licensed. GMRS radios offer more power than FRS, allow external antennas, and (best of all) cover your whole household under a single FCC license.

Contingency is all about options that extend your range and flexibility when the "everyday" tools aren't cutting it.

Emergency – Last-Ditch Signals

Let's say it all goes wrong. No cell service. No radios. No power. No solar backup. You're down to caveman comms. Don't laugh—this stuff still works.

Your emergency layer should include:

Whistles or air horns – loud, carry far, and require no power.

Signal mirrors – sunlight flashes can be seen for miles.

Flares or flashlights – for nighttime signals.

Smoke or signal fires – a classic, but still effective.

These aren't your go-to's for daily use. They're the "break glass in case of emergency" tools. Save them for when everything else has failed, because they'll make you visible—but they'll also let others know exactly where you are.

Why P.A.C.E. Matters for Preppers

Here's the big takeaway: redundancy isn't overkill—it's survival. A smartphone alone isn't a comms plan. Neither is a Baofeng in a drawer. Layer your tools, train with them, and know exactly when to fall back to the next option.

That way, when disaster strikes, you're not standing there staring at one dead gadget—you've already got Plan B, C, and D ready to roll.

Wilderness Protocol: Is It Really Necessary?

The short answer: yes.

The Wilderness Protocol is one of those little gems in the radio world that most people never hear about until they need it—and by then, it's too late. At its core, the Wilderness Protocol is simply a recommendation: when you're outside of repeater range, monitor standard simplex calling frequencies at specific times so that anyone in need has a fighting chance of being heard.

This idea has been around for decades—originally mentioned in QST Magazine way back in June 1996—and it's still just as relevant today.

Why It Matters

Now that summer's here, a lot of us are getting outdoors: camping, hiking, off-roading, fishing. For preppers, that means we're often in the backwoods or mountains—exactly the places where cell phones fail and repeaters don't reach.

That's where the Wilderness Protocol shines. By knowing the calling frequencies and checking in at the recommended times, you create a kind of "safety net" in the wilderness. If someone gets hurt, stranded, or lost, they know there's a chance someone else is listening. And sometimes, that chance is all it takes to save a life.

Frequencies to Monitor

Here are the recommended simplex calling frequencies for Amateur Radio:

146.520 MHz – 2 meters (the national VHF calling frequency, primary)

446.000 MHz – 70 cm

223.500 MHz – 1.25 meters

52.525 MHz – 6 meters

1294.500 MHz – 23 cm

For other services:

GMRS: 462.675 MHz (Channel 20, tone 141.3) – National Emergency Calling Frequency

CB: Channel 9 – monitored by many law enforcement agencies (use for emergencies only)

When to Monitor

To conserve battery life, the Wilderness Protocol recommends monitoring at least five minutes at the top of the hour:

07:00 – 07:05

10:00 – 10:05

13:00 – 13:05

16:00 – 16:05

...and so on.

If possible, listen every hour on the hour.

Priority/Emergency Calls: Start with 10 seconds of DTMF "0" (called LiTZ—Long Tone Zero). It cuts through chatter and grabs attention.

Routine Calls: Wait until four minutes past the hour before calling, so emergencies get priority.

How It Works in Practice

Let's say you're hiking in the Rockies and twist your ankle miles from the trailhead. You're out of cell range, no repeaters in sight. You switch to 146.520 MHz, wait for the top of the hour, and call for help. Someone in a valley or even another hiker miles away might be monitoring—and that might be your lifeline.

Flip the scenario. You're at home in your radio shack and hear a weak, scratchy signal calling for help on the national calling frequency. You may be the only one who hears it. That makes you the critical link—gather their info, get their location if possible, and immediately relay it to local search and rescue.

Why Preppers Should Care

For us, the Wilderness Protocol isn't just about backcountry adventures—it's another tool in the communications toolkit. Coupled with your P.A.C.E. Plan, it gives you layered resilience:

PACE: Your structured redundancy (Primary, Alternate, Contingency, Emergency).

Wilderness Protocol: Your wilderness safety net when the usual layers are out of reach.

Put the two together, and you'll have peace of mind knowing you've prepared for the "what ifs" that so many people ignore until it's too late.

Bottom line? The Wilderness Protocol costs you nothing but a few minutes of listening. And if you ever find yourself—or someone else—in trouble, those few minutes could make all the difference.

CTCSS Codes: What They Are and Why They Matter

If you've ever tried using a radio and wondered why you can't hear anything—or why people can't hear you—you may have bumped into CTCSS codes.

CTCSS stands for Continuous Tone-Coded Squelch System. Don't let the technical name scare you; it's basically a way for radios to filter out unwanted chatter.

Think of it like having a secret knock on a door. The door isn't locked—anyone can still open it—but unless you use the right knock, the person inside won't answer.

How It Works

When you transmit with a CTCSS code, your radio adds a low-frequency tone (between 67 Hz and 254 Hz) under your voice signal.

The other person's radio is set to listen for that same tone.

If the tone matches, their squelch opens and they hear you.

If it doesn't match, their radio stays quiet—even though your signal is technically there.

Important: CTCSS does not make your conversation private. Anyone without the code can still hear you if they set their radio to "monitor mode" or disable tone squelch. What it does is reduce background noise and make sure you only hear your people.

Why Preppers Should Care

In a real-world scenario:

Without CTCSS: You hear every kid down the street with a Walmart walkie-talkie and every random conversation on the same channel.

With CTCSS: You only hear transmissions from your group, because you're all using the same tone.

This keeps your radio quieter, conserves battery (since you're not constantly transmitting "noise"), and reduces confusion when multiple groups are active nearby.

Common Use Cases

GMRS and FRS Radios – Most blister-pack walkies have CTCSS built in. They might call it "privacy codes" or "sub-channels." (Spoiler: it's not real privacy, just tone squelch.)

Repeaters – Many Amateur and GMRS repeaters require a specific CTCSS tone to access them. If you don't send the right tone, the repeater won't retransmit your signal.

Group Coordination – MAGs (Mutual Assistance Groups) can assign specific tones for different teams or tasks.

Examples of CTCSS in Action

You and your group agree to use Channel 16, Tone 123.0 Hz on GMRS. Everyone sets their radios the same way. Now your radios stay quiet until one of your group keys up.

A local GMRS repeater lists its frequency as 462.725 MHz, +5 MHz offset, Tone 141.3 Hz. Unless you program in that 141.3 Hz tone, you'll never open the repeater.

On ham radio, your club's repeater might say: "146.940 MHz, -600 kHz offset, PL 100.0." That "PL" is Motorola's old brand name for—you guessed it—CTCSS.

Quick Reference: Common Tones

Some of the most frequently used tones include:

67.0 Hz

88.5 Hz

100.0 Hz

123.0 Hz

141.3 Hz

156.7 Hz

Most radios give you a menu of 38 or more standard tones to choose from.

Pro Tips

Always write down your group's frequency + tone in your comms plan. Example: "GMRS Ch20, 462.675 MHz, Tone 141.3."

Test your setup before you need it. Mismatched tones are one of the most common reasons groups can't hear each other.

Don't rely on tones for security. Anyone with a scanner or radio set to monitor mode can hear you.

Bottom line: CTCSS is a simple but powerful tool for keeping your radios quiet and your group organized. Use it to cut through the chaos—but don't mistake it for encryption or secrecy. It's about clarity, not privacy.

#1000NewHams

A new year always brings new opportunities. Fresh starts, new projects, and—for those of us in the Amateur Radio and prepping worlds—a renewed call to action. That's where #1000NewHams comes in: the push to encourage at least a thousand new people to get their ham license and join the community.

Before I talk about why this matters, let's rewind a bit.

My Journey into Prepping (and Radio)

Like many of you, I came of age during the Y2K scare. Remember that? We were told the world's computer systems would roll over to "00" and civilization would collapse. Spoiler alert: it didn't happen. But that experience taught me something important—while the world didn't end, there were plenty of smaller, everyday disasters worth preparing for.

So I shifted my mindset. Instead of prepping for the "end of the world as we know it," I prepped for the things that actually happen:

Severe weather

Job loss

Extended illness

Power outages

Real-world problems, with real-world solutions. And for the most part, I did well—food, water, shelter, medical. But there was one big hole in my preps I didn't even notice at first: communications.

Sure, I had a TV and a radio like everyone else, but I didn't have a way to reach out when the grid went down. That changed around 2008 when I discovered Amateur Radio. By 2009, I had passed my Technician exam. Ten years later, I earned my General license. My first HF rig was a little single-band radio, and I still use it today for portable comms.

Why Amateur Radio Belongs in the Top 3

Preppers talk about the "big three": food, water, and shelter. I'd argue we need to make it the big four: food, water, shelter, and communications.

Here's why:

Humans are social. We don't just survive alone—we thrive when we can connect with others.

Information saves lives. Weather alerts, road conditions, supply updates, medical requests—all flow faster when you have working comms.

Mental health matters. In a long-term grid-down event, just hearing another human voice can help keep you grounded and sane.

Amateur Radio delivers on all three.

A Few Ground Rules

Now, before you go out and buy a Baofeng, you need to know a few things:

Amateur Radio is a licensed service. You need an FCC license grant to transmit. Listening is legal, transmitting without a license is not.

Handheld radios (HTs) have limits. On simplex (radio-to-radio), you'll usually get about a mile or so. With a repeater, you might stretch that to 30–50 miles.

Repeaters are key. They sit on mountaintops and tall buildings, listening for your little 5-watt signal and rebroadcasting it at higher power. That's how you cover more ground with a small handheld.

If you don't know how to program your radio for a repeater, it's just a $20 paperweight. Thankfully, there are YouTube channels, blogs, podcasts, and step-by-step guides that can walk you through it.

Going Beyond the Handheld

Handhelds are a great starting point, but sooner or later, you'll want more reach. The good news is that even Technician licensees have voice privileges on the 10-meter band (28.300–29.300 MHz). With the right antenna, that opens the door to worldwide communication—no repeater required.

Why #1000NewHams Matters

Here's the heart of it: Amateur Radio isn't just a hobby—it's a preparedness skill. Every new ham adds resilience to their family, their group, and their community.

Think about it:

A thousand new hams = a thousand new radios, a thousand new sets of ears, and a thousand new voices in the network.

That's a thousand more people who can pass emergency traffic, check in on neighbors, or call for help when cell towers are dark.

That's a thousand more people who keep the spectrum alive, active, and useful.

If you're on the fence, this is your year.

How to Get Started

Step 1: Start studying with free tools like HamStudy.org, QRZ.com, or YouTube crash courses.

Step 2: Take practice exams until you're scoring above 80% consistently.

Step 3: Find an exam session near you—or online through the ARRL. Cost is usually around $15.

Step 4: Once you pass, buy a basic dual-band handheld, program your local repeaters, and start practicing.

It's not as hard—or as expensive—as you might think.

Final Word

Prepping isn't just about beans, bullets, and band-aids. It's about being ready to adapt, survive, and thrive no matter what comes your way. Communications is part of that.

So here's my challenge to you: let's make this the year of #1000NewHams. Get your license, get on the air, and bring someone else along with you. The more voices we add to the network, the stronger and more resilient we all become.

Who knows? Maybe I'll hear your call sign on the air one day. Until then—73, my friends.

Appendix A

Comms Plan Template

Prepper Communications Plan Template

Group / MAG Name: _____

Date Created: _____

Last Updated: _____

1. Contact List

(Every group member should have this same list.)

Name / Call Sign: _____

Primary Radio Type (HAM, CB, FRS, GMRS, MURS): _____

Primary Frequency / Channel: _____

Secondary Frequency / Channel: _____

Cell / Text Number (if available): _____

Other Notes (skills, roles, medical needs, etc.): _____

2. Frequencies & Channels

Service Channel / Frequency Notes (Tone, Shift, etc.) Primary / Backup

HAM 2m _____ _____ ☐
Primary ☐ Backup

HAM 70cm _____ _____
☐ Primary ☐ Backup

CB _____ _____ ☐
Primary ☐ Backup

FRS _____ _____ ☐
Primary ☐ Backup

GMRS _____ _____ ☐
Primary ☐ Backup

MURS _____ _____ ☐
Primary ☐ Backup

Other _____ _____ ☐
Primary ☐ Backup

3. Check-In Schedule

Top of the Hour: ☐ Yes ☐ No

Every ____ Minutes: _____

Daily Roll Call (Time): _____

Emergency Only (Break Glass Rule): ☐ Yes ☐ No

4. Call Signs & Procedures

Group Call Sign (if used): _____

Individual Call Signs: _____

Brevity Codes / Signals:

"All Clear" = _____

"Need Help" = _____

"Move to Secondary Frequency" = _____

5. Backup Methods

Text / SMS Tree: _____

Designated Relay Station (Name / Location):
_____

Alternate Meeting Point if Comms Fail:
_____

6. Practice Schedule

Weekly Radio Net (Day / Time): _____

Monthly Full Drill (Date): _____

Last Practice Completed: _____

Appendix B

Getting Started in Ham Radio: A Step-By-Step Guide

So you've decided Amateur Radio might be worth your time. Good call. Now let's break down the process into bite-sized steps so it doesn't feel overwhelming.

Step 1: Learn the Basics

Don't panic—you don't need to be an engineer. The entry-level Technician license test is 35 multiple-choice questions. Topics include:

Basic radio theory (how signals travel)

Safety (like not zapping yourself with RF)

FCC rules and operating practices

Most folks can study for it in a few weeks of casual effort.

Step 2: Pick a Study Resource

You've got options:

HamStudy.org (free, flashcards + practice tests)

Ham Radio Prep (structured courses, good for beginners)

ARRL Exam Review (official study site)

YouTube – search "Technician ham radio crash course" for video lessons

Books – The ARRL Ham Radio License Manual is the gold standard

Tip: Take practice exams until you're consistently scoring above 80%. At that point, you're ready.

Step 3: Find an Exam Session

Exams are given by teams of Volunteer Examiners (VEs). You can take the test two ways:

In person: Usually hosted by ham radio clubs, often once a month.

Online: Thanks to Zoom and webcams, you can now test from home.

Find sessions at:

http://www.arrl.org/find-an-amateur-radio-license-exam-session

http://www.arrl.org/online-exam-session

Cost: usually $15 or less.

Step 4: Pass the Test

When you pass, the VE team sends your info to the FCC. Within about a week, your callsign will show up in the FCC database. That's your official green light to transmit. Print it, screenshot it, tattoo it (okay, maybe not that last one).

Congratulations—you're now a ham.

Step 5: Get Your First Radio

Start small, start simple:

Handheld (HT): A Baofeng UV-5R (about $25) is cheap and works. Not the Cadillac of radios, but great for learning.

Upgrade Antenna: Toss the stock "rubber duck" antenna and get an aftermarket dual-band whip. Huge difference in performance.

Accessories: Extra batteries, a simple programming cable, and CHIRP software (free) will make your life easier.

If you've got the budget, consider a mobile radio (25–50 watts) with a mag-mount antenna for your vehicle or even as a small base station at home.

Step 6: Program and Practice

Learn how to program your local repeaters into your radio (RepeaterBook.com is your friend).

Get on the air—listen first, then key up when you're confident.

Join a local net to practice. Nets are like on-air meetings; they're structured, friendly, and designed for exactly this purpose.

Step 7: Level Up

Once you're comfortable as a Technician:

Study for your General license (another 35 questions). This opens up HF bands and worldwide communication.

Eventually, shoot for Extra if you want full privileges.

Step 8: Connect with the Community

Ham radio is best learned with others. Join a local club, attend field days, and meet people who can help you grow. Most hams love teaching newcomers—it's part of the culture.

Step 9: Integrate Into Your Preps

Finally, make Amateur Radio part of your overall plan:

Write your frequencies into your Comms Plan.

Train your family or MAG on how to use the radios.

Run regular drills.

Keep backup power (extra batteries, solar chargers) handy.

Radios aren't "buy once and forget" gear. They're tools you sharpen through use.

Bottom Line

Getting into ham radio isn't as hard—or as expensive—as it used to be. For the cost of a tank of gas and a couple evenings of study, you can earn a license, buy a radio, and plug into a worldwide network of operators. More importantly, you'll give yourself and your group a reliable way to communicate when everything else goes dark.

Appendix C

P.A.C.E. Plan Worksheet

Group _____ / MAG _____ Name: _____

Date Created: _____ Last Updated: _____

Primary (Everyday Comms)

Your first choice for communication when everything is normal.

Device/Method: _____

Phone Numbers / Text Tree:
_____

Backup Power Source (power bank, solar, vehicle charger):
_____

Alternate (Go-To Backup)

What you fall back on if your primary fails.

Device/Radio Type (HAM, GMRS, FRS, CB, etc.):
_____

Frequencies / Channels:
_____

Range / Limitations:
_____

Notes (repeaters, antenna setup, licensing info):
_____

Contingency (Specialized Tools)

Used when both Primary and Alternate are unavailable or compromised.

System/Method (APRS, GMRS, mesh network, etc.):
_____

Equipment Needed:
_____

Coverage Area: _____

Who in the group has this capability?
_____

Emergency (Last-Ditch Signals)

When everything else is down, these are your final methods of signaling for help or passing info.

Visual (flares, mirrors, flashlights, smoke):
_____

Audible (whistle, horn, banging metal):
_____

Other: _____

Group Signal Codes (e.g., 3 whistle blasts = emergency):
_____

Practice Schedule

Drill Dates: _____

Last Practice Completed:
_____

Next Practice Planned:
_____

Tip: Print one copy for each group member. Fill it in together, keep it in waterproof sleeves, and stash extras in your go-bags. Radios fail, phones die—but a written P.A.C.E. plan will still be there when you need it.

Appendix D

Wilderness Protocol Quick Reference

Purpose: Provide a safety net for radio operators outside of repeater or cell coverage. Monitor key frequencies at specific times so emergencies can be heard and relayed.

Amateur Radio Calling Frequencies (Simplex)

146.520 MHz – 2 meters (Primary)

446.000 MHz – 70 cm

223.500 MHz – 1.25 meters

52.525 MHz – 6 meters

1294.500 MHz – 23 cm

Other Services

GMRS: 462.675 MHz (Channel 20, tone 141.3) – National Emergency Frequency

CB: Channel 9 – Emergency / monitored by law enforcement

Monitoring Times

Listen at least 5 minutes at the top of the hour:

07:00 – 07:05

10:00 – 10:05

13:00 – 13:05

16:00 – 16:05

(and so on every three hours — or every hour if possible)

How to Transmit

Emergency / Priority:

Begin with 10 seconds of DTMF "0" (LiTZ – Long Tone Zero).

Clearly state: "This is an emergency."

Give location, nature of emergency, and assistance needed.

Routine Calls:

Wait until 4 minutes past the hour to avoid blocking emergencies.

Example: "This is [Call Sign/Name] monitoring [Frequency]."

Field Tips

☑ Conserve battery—keep radio off except during check-in times.

☑ Use highest ground available for best line-of-sight.

☑ Carry backup power (extra batteries, solar, power bank).

☑ Keep comms simple, clear, and brief.

Remember: You may be the one calling for help—or the one who hears it. Either way, the Wilderness Protocol ensures someone is listening.

Appendix E

CTCSS Cheat Sheet

What it is:

CTCSS (Continuous Tone-Coded Squelch System) uses sub-audible tones (67–254 Hz) to reduce unwanted chatter and access repeaters. Radios must match on both frequency and tone to hear each other.

Standard CTCSS Tones

Tone (Hz)	Common Uses	Notes
67.0	GMRS repeaters, FRS channels	Often used for rural repeaters
77.0	GMRS / business radios	Less common, but available
88.5	Amateur 2m repeaters	Popular "quiet tone"
100.0	Ham & GMRS	Very widely used
103.5	GMRS & public service	
110.9	Amateur repeaters	
123.0	GMRS (family groups)	Easy to remember
131.8	Amateur & commercial systems	
136.5	GMRS, MURS	
141.3	National GMRS emergency calling frequency (Ch 20)	One every prepper should know
146.2	GMRS / ham	

156.7 Amateur repeaters Common "default" tone

162.2 GMRS

167.9 Ham / GMRS

173.8 GMRS

179.9 Ham repeaters

186.2 GMRS

192.8 Amateur

203.5 Ham repeaters 210.7 GMRS 218.1 Amateur repeaters

225.7 GMRS 233.6 Amateur 241.8 GMRS

250.3 GMRS / ham Upper end of standard tones Quick Tips for Preppers

• Always pair tone + frequency. Example: "GMRS 462.675 MHz, Tone 141.3 Hz"

• Same channel, different tone = no contact. Both radios must match.

• CTCSS ≠ privacy. Others can still listen in with "monitor mode."

• Write it down. Put tones in your Comms Plan and P.A.C.E. worksheet so everyone in your group is on the same page.

Appendix F

#1000 NewHams Quick Start Guide
Your path to becoming a licensed ham radio operator

◆ Why Ham Radio?

Works when cell towers and internet don't.

Local, regional, and worldwide communication.

Strengthens your family, MAG, and community preps.

Builds skills, confidence, and resilience.

◆ Step 1: Learn the Basics

☐ Amateur Radio is a licensed service (FCC required to transmit).

☐ Three license levels: Technician → General → Extra.

☐ Start with the Technician exam (35 multiple-choice questions).

◆ Step 2: Study for the Test

☐ Free Online: HamStudy.org

, QRZ.com

☐ Books & Courses: ARRL License Manual, Ham Radio Prep

☐ YouTube Crash Courses: search "Technician Ham Radio"

Tip: Practice until you score 80%+ consistently.

◆ Step 3: Take the Exam

- ☐ In-person with a local club or online via Zoom.
- ☐ Cost: Usually $15 or less.
- ☐ Find sessions:

ARRL Find a Session

ARRL Online Exams

◈ Step 4: Get Your First Radio

- ☐ Dual-band handheld (HT) – e.g., Baofeng UV-5R (~$25)
- ☐ Upgrade antenna (stock antennas are weak)
- ☐ Learn to program local repeaters (RepeaterBook.com)
- ☐ Accessories: extra batteries, programming cable, CHIRP software (free)

◈ Step 5: Practice & Connect

- ☐ Listen first, then try 146.520 MHz (2m calling frequency)
- ☐ Join a local net to practice structured comms
- ☐ Connect with experienced hams—most love to help new ops

◈ Step 6: Grow Your Skills

- ☐ Upgrade to General Class for HF (nationwide/worldwide)
- ☐ Try digital modes (APRS, Winlink, FT8)
- ☐ Add ham radio to your P.A.C.E. Plan

Your Mission:

Get your license. Get on the air. Bring someone with you.

Let's make this the year of #1000NewHams.

The more of us on the air, the stronger we all are.

Final Thoughts

Preparedness isn't about paranoia—it's about peace of mind. It's about knowing that when the lights go out, the storms roll in, or the unexpected happens, you have the tools, the knowledge, and the confidence to keep moving forward.

Radio communications are a big part of that picture. They connect us not only to information, but to each other. Whether you're reaching a neighbor across town, a group across the county, or a stranger across the globe, you're proving the point: we're stronger when we can communicate.

So practice. Build your plan. Run your drills. And when the time comes, you'll be ready to key that mic and say the words that matter most.

About the Author

M.Ray Davis has been a prepper since 1998, when the Y2K scare first sparked his interest in self-reliance. While the world didn't end that New Year's Eve, he quickly realized that real-world events—storms, power outages, job loss, illness—were more than enough reason to build a preparedness lifestyle.

By 2009, he had earned his Amateur Radio license, discovering what would become a lifelong passion for radio communications. A decade later, he upgraded to General class, opening the door to worldwide contacts and the ability to connect with people across the globe.

A true believer in practical prepping, M.Ray Davis has spent more than two decades combining food, water, shelter, and communications into a sustainable lifestyle—without panic, without hype. His mission is to help others do the same: prepare for the everyday disruptions that really happen, and build the skills to face them with confidence.

When he's not on the air or testing gear in the field, you'll find him writing, podcasting, and teaching others how to step into preparedness with common sense and a little bit of humor.

www.ingramcontent.com/pod-product-compliance
Lightning Source LLC
Chambersburg PA
CBHW032010080426
42735CB00007B/556